The

Paranormal

~

In search of answers

By

Rich Arrington

Introduction

In the pages of this book, I will explain my journey from childhood to career law enforcement officer and on to a psychic paranormal investigator.

I define myself as a psychic paranormal investigator because the impressions I get come from my clairvoyant ability and my abilities have been tested to 88% and 98% accuracy. On my website I have a saying, "I see what I see and I see nothing more." This is because a true clairvoyant can't control what images come to them. Some images are not understood by the clairvoyant themselves and it must be determined by those requesting the information if it has meaning.

I'll explain some of my own beliefs in psychic ability and about other paranormal activity in the pages of this book.

Table of Contents

The Early Years

East Tennessee is a wonderful place to grow up and as a child; I can recall spending weekends with my grandmother and helping her can vegetables. You learn a lot about the "old ways" or your heritage, growing up in this area.

Today, I take my own son's to Historic sites in East Tennessee. They like the Exchange Place on Orebank Road in Kingsport, Tennessee. At the Exchange Place, my son's learn about their heritage. They also learn about native crafts from the region and this helps them to develop a pride in their East Tennessee heritage.

I had a rather normal childhood growing up in East Tennessee. My mother was a homemaker and my father was a police officer. I grew up in a large family with four brothers and two sisters. Our house was very busy on any given day.

With my own two small sons, ages 11 and 5, I know about the noise and turmoil that two kids can bring to a house on a Saturday morning. My two sons' can cause a major disturbance over who gets the last cookie or which Saturday morning cartoon is going to be watched. I sometimes wonder how my parents dealt with seven children in the house.

I don't recall many major events in our family, things seemed rather normal even for the early 1970's. I recall the "Peace Movement" and my brothers and sisters rock music and odd styles. Now, I think back to those days with amusement and how much my brothers and sisters have changed over the years. The middle aged adults I meet today and call my brothers and sisters are nothing like the hippies

I grew up with. We are still very close and love each other very much.

My grandfather on my Mother's side was a Faith Healer and a farmer. My grandmother was a country woman who dipped snuff and was a good foot taller than my grandfather.

My grandfather was what at the time some called a "prophet." I am not sure why people used that term. I guess because the term "psychic healer" was not understood or they found disturbing for some reasons. My grandfather would often be visited by people with illnesses and they would request healing by prayer.

My grandfather was once visited by a mother whose daughter was having "nightmares" and "outbursts" of cursing and the mother was in fear of her daughter. It was reported this behavior had come upon the girl sudden and without warning. My grandfather visited the home of this concerned mother. He found the daughter who appeared to be about twelve years of age sitting in the corner of her bedroom and screaming at unseen images.

My grandfather stood in the center of the child's room. She appeared to be yelling and cursing at a person at the far end of the room. My grandfather could not see anyone else in the room but felt a presence. He called to the girl to get her attention. The girl only responded to him after the third time he called her name.

My grandfather could communicate with the child but she appeared to him to be under the influence of dark forces. My grandfather could see shadows moving along the walls and this appeared to be who the girl was communicating with.

My grandfather began to read Psalm 91 from the Bible he carried and placed his hand on the girl's head who jerked away in protest. He continued to pray and read for hours (the same verse from Psalm 91) until the girl's emotions turned from anger to tears. He wrote the verse from Psalm 91 on a piece of paper and placed it between the mattress of the girl's bed and put her back in the bed due to her own exhaustion from the ordeal. The girl went to sleep and after she awoke the next day, her mother reported that the outbursts had stopped and they also never returned. The verse from Psalm 91 is something my grandfather told me as a child, I'm not sure why but I feel its something I should keep to myself. Maybe it's because I will need to use it in some event in the future?

Was this a 1930's version of a southern Appalachian exorcism?

My First Encounter

In 1972, I was eight years old, my dad was an investigator and we moved into a farmhouse near Greeneville, Tennessee. The house was fun for me at first; I loved the farm life with chickens, pigs, cows and the large barn behind the house. Daily I would go to the barn to explore the loft for old objects. Once I found a lantern that I later used for camping with my brothers.

On one Saturday, my sister and I were playing in the barn, all seemed normal and we went up into the loft to get a better view of the property. We opened the loft window and sat looking out over the property.

As we sat, I could feel we were not alone and a few times I could feel someone touch me on the back. I told my sister and she laughed as often children do in these situations.

I felt the touch repeatedly and began to get frightened and told my sister I wanted to leave the barn. She assured me it was nothing, so we continued to sit. I felt someone was behind me and I turned but no one was there. A few minutes past and I turned a second time and came face to face with a man who appeared to have no eyes but he was smiling at me. I could see dark holes where his eyes should've been. At this point, I was unable to speak and I turned and began to walk backwards and I fell from the loft hitting the floor below. My sister jumped to grab me and prevent the fall but it was too late.

I hit the bottom and the fall knocked the breath out of me. My sister climbed down the ladder leading from loft and proceeded to check on me. All I could do was point up at the loft and utter the word, 'man.' My sister looked up and said, "What man?" she had not seen the man in the loft.

My sister's yelled to summon my dad, who was working near our chicken coop. He checked on me and then went into the loft. He came back down to say he could not find a man in the loft. Was this the end of the story? No!

 The month's that followed got worse, much worse. My dad became locked in the chicken coop repeatedly and my four brothers' saw the "man" in the living room and on one occasion he walked through a wall.

   My mother saw the "man" in the hallway numerous times and it was always followed by a puddle of urine in the spot where the man was standing.

   My grandmother, who was visiting, saw the "man" standing near the barn and looking at the house. My sister's saw the "man" in their bedrooms in the middle of the night.

After a few months, the situation at our Greeneville, Tennessee home got to the point none of us felt safe. The ghost or haunting became worse over time. We moved out and never looked back.

Premonitions

Later, in my adult life, I followed my father's career and became a Deputy Sheriff. I never really liked the job, but found it to be a career I already knew and was ready for and familiar with.

I guess a lot of people tend to follow the career path of their parents; it seems the easy thing to do. I later found this to be the difficult thing for me to do.

Why? This is because I found during my career that I had Clairvoyant ability and the premonitions were sometimes very difficult to see and deal with.

Premonitions or precognition is the direct knowledge or perception of the future, obtained through extrasensory means.

Precognition is the most frequently reported of all extrasensory perception (ESP) experiences and one I've experienced most often, occurring in my concentration on an object. It may also occur spontaneously in visions of future events, flashing thoughts entering the mind, and the sense of "knowing." Precognitive knowledge also may be induced through meditation.

Usually the majority of precognitive experiences happen within days prior to the future event, and sometimes its within a few hours. Precognitive experiences can occur months or even years before the actual event takes place.

My experience with precognition has brought on some stressful events with seeing disturbing images of things to come. I've seen the future in relation to numerous deaths of friends and relatives that have occurred over time. Not all has been bad, I've seen events such as one fellow officer who was going on a call and I gave him a warning, thankfully he believed me and was able to prevent a suspect from harming him during the call. Later my friend bought my lunch as payment for the warning.

Not everyone believes in precognitive ability which is OK, but I've found from my own experience that it's foolish to think it does not exist.

In 2002, I got the urge to sketch a farm house (see image 1) and did not understand why. A vision of a farm house came into focus. I sketch the room locations and a trail leading from the driveway of the house to a larger home. I gave the sketch to my wife and ask her to keep it in a safe place. Having numerous visions as I do, I forgot about the sketch until a few months later and was approach by a friend who had a house for us to look at, a few days later we were taken to the house in the sketch.

13

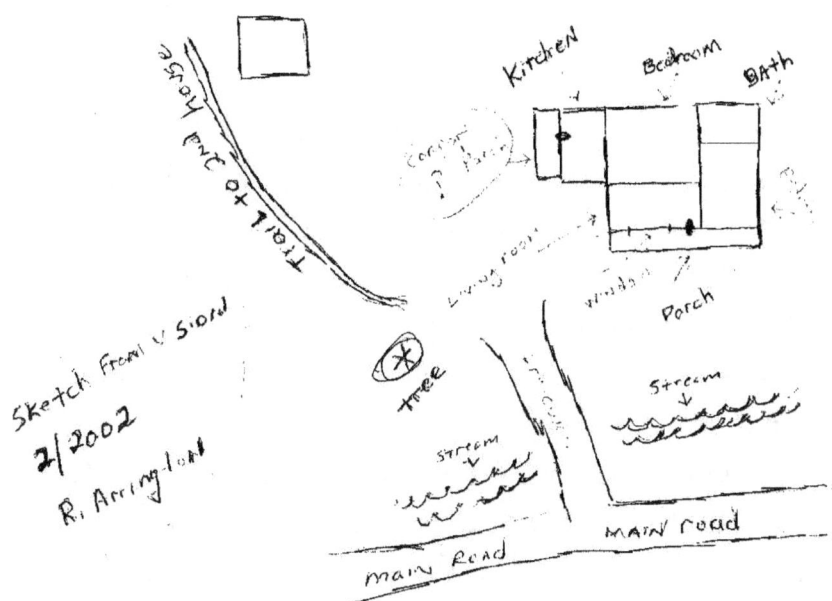

Image (1)

14

My wife and I were somewhat surprised to find the house was so much like the sketch I had drawn months before and we told the owner we would like to see the trail leading from the house. He was a little surprised by the request and could not understand how we knew of the trail. After making an excuse as to why we had knowledge of the trail, he agreed to show it to use and explained that the trail lead to his mother's old house. Today, we live in the house at the end of the trail. We lived in the original house in the first sketch until renovations was complete on the second house. In the second sketch (see image 2) we found that only the tree near the drive was different and the location of the stream.

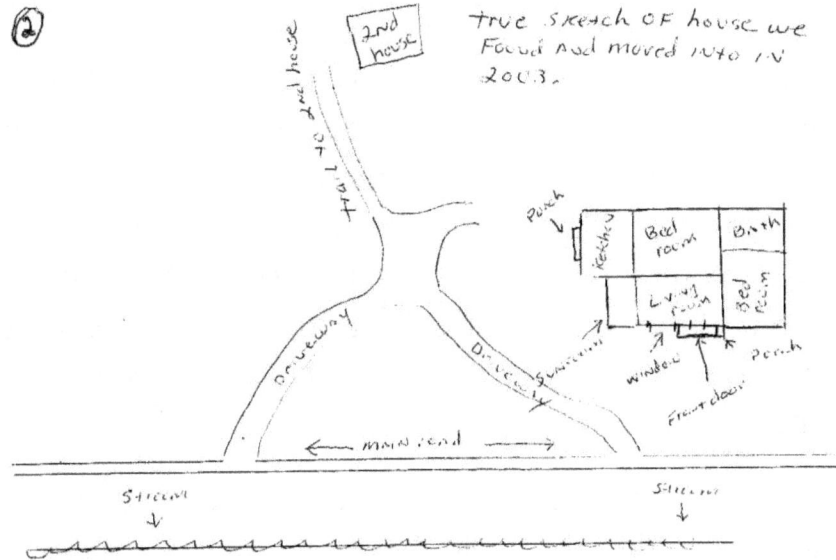

Image (2)

16

Dreams

On my radio show I often have callers who are disturbed by dreams and ask if the images in these dreams have meaning. I can only answer, yes. Some images in dreams do have meaning and others are nothing more then memories replying themselves over in our subconscious. As a clairvoyant, I have precognitive dreams often that foretell future events and some are visions of past events or postcognative dreams.

Postcognition or Retro cognition is the ability to achieve information about events that has already occurred which are usually unknown by others. Clairvoyants use this method to detect possible murderers and missing persons. I often used this ability in my law enforcement career and still use it today in missing person cases and to see events in hauntings.

In 1974, I was sitting in the back seat of my parents station wagon at a McDonald's in Asheville N.C. As we were eating, I ask my mother where the "Rock Building" was at in Erwin, Tennessee. "It's on North Main Street, why?" she asks. "I was just thinking about the time when we took grandpa his lunch, when he was doing the rock work." I said. I went on to describe my dad's truck and what my mother was wearing and where my grandpa ate his meal.

My grandpa built the "Rock Building" in 1947, the year my parents were married and seventeen years before I was born. This odd event in my parents station wagon is something I will never forget because it was the day my parents realized and so did I that I had clairvoyant abilities. The real odd part of this story is that I can still see the events surrounding the 1947 construction of the rock building as if I was truly there at the time.

The vision of the rock structure was not a dream but it's the best example of postcognative ability I can share in this book. In most dreams, one tends to have precognitive visions and I had one of these in 1990 prior to meeting my wife. I was taking a nap on my sofa and dreamed of a red headed girl holding a baby. The vision (dream) was in front of a brick house and the girl was sitting on a porch swing to the left of the front door. I came out the front door of this house and looked to the left to hear this girl say "there's daddy" as she smiled.

One year later I met this girl at a party and I recognized her from my dream. We were married a year later. In 2001, I found myself exiting the front door of this same brick house and saw my wife hold up our son and say "there's daddy." So, do dreams come true? Yes!

ESP

ESP is not a term you hear much any more, that is because psychic ability is more defined now days and you tend to hear the terms, telepathy or clairvoyance. Extrasensory perception (ESP) involves reception of information not gained through the recognized physical senses but sensed with the mind. The term denotes psychic abilities such as telepathy, clairvoyance, and precognition. ESP is also sometimes casually referred to as a sixth sense.

Extrasensory perception is said to occur spontaneously in conditions which are not scientifically controlled. Such experiences have often been reported to be much stronger and more obvious than those observed in laboratory experiments.

These reports, rather than laboratory evidence, have historically been the basis for the widespread belief in the authenticity of these phenomena. However, it has proven extremely difficult to replicate such extraordinary experiences under controlled scientific conditions. Most people with ESP can't control the phenomenon or predict its occurrence.

Events come to the person through visions or dreams and not by choice. I never have believed that there is a single test to determine ESP and even though I have allowed myself to be tested four times with a test score of 88% to 98% accuracy in my predictions, I do not believe these tests to be a single factor in determining ones ability.

Clairvoyance is a form of ESP and often used to refer to other forms of anomalous cognition, most commonly the perception of events that have occurred in the past, or which will occur in the future known as postcognition or retro cognition and precognition respectively.

Clairvoyance is related to remote viewing, although the term "remote viewing" itself is not as widely applicable to clairvoyance because it refers to a specific controlled process. This form of ESP is what I discussed in the last two chapters and I think you now see that they are connected.

Hauntings

There is no single haunting that is a like because ghosts are people to and have their own personalities. There are also different kinds of hauntings and here is a list:

Residual haunting activity can occur when something traumatic occurs, such as a murder. Negative energy is literally in the atmosphere, causing the atmosphere to record the events. The entities involved in this residual haunting activity are unaware of their surroundings. There is no interaction between you and the entity.

Poltergeist activity is usually caused by an adolescent teenage girl going through puberty stages. During this stage of development, a young girl can harbor an extreme amount of inner energy. This energy can be projected with the mind, which can cause movement of objects. Poltergeist activity usually originates from a human being.

The Brown Lady, one of the earliest photo's of a spirit.

A photo sent to me in 2011 of a reported ghostly janitor.

Demons are entities that never had a mortal human form. To handle demonic cases, it is essential that a blessing of the purest kind is conducted on the person or home. Consult your clergyman, priest or Demonologist for help with this kind of activity, do not allow an inexperienced person to handle this most dangerous form of paranormal activity.

Intelligent/interactive haunting

When our mortal form dies, the aura that constantly surrounds our bodies leaves our bodies. This energy, the aura, or soul, is carrying the information of what we used to be. If it can do this, then why couldn't it also carry our intelligence? If it can carry our former intelligence of our previous life, then it should be able to interact with us intelligently. When we see this aura or soul, we call it a ghost. If this ghost is able to interact with us, is aware of us, can touch us, can communicate with us, then this is an intelligent/interactive haunting.

Shadow figures/Shadow people

This is a type of haunting activity that has no real explanation. They are different from ghosts. They are usually shapeless dark masses. They are known to do things that are different from ghosts. They can move between walls, they have no human features and they are sometimes hooded shadow creatures. Those who encounter them, have a feeling of dread. Shadow People have no discernible mouth, noses or facial expressions. There are also reports of shadow animals, such as a shadow in the form of a cat, with no discernible mouth, nose or eyes. I have felt for some time that some of these figures are also demonic and I classify them as dangerous and would advise caution in dealing with such entities.

This image of a shadow figure appeared on a camera in the middle of a room. The people there reported a feeling of anxiety and dread. They also reported high pressure associated with a demonic entity. A shadow figure can be misunderstood by many and from my own experience I know it can cause emotional feelings from those present in a room. If you see a shadow figure, call an experience paranormal team to deal with the entity.

Haunted People

Do you here things in the night? Are you awakened by voices or sounds somewhere in your home? Do you see unexplained mists, shadows images that rush along your walls or ceiling in the night? Perhaps you have seen things that looked out of place, and felt it may be a ghost?

A ghost is a human being who has passed out of the physical body, usually in a traumatic state and is not aware usually of his true condition. We are all spirits encased in a physical body. At the time of passing, our spirit continues into the next dimension. A ghost, on the other hand, due to trauma, is stuck in our physical world and needs to be released to move on.

You may have the kind of ghost that's said to be attached to a person and is sometimes earth bound for a short time after their death. It is said they stay close to their loved ones. They are very aware they have died, but stay for a while to comfort the people they left behind. It is said they have the power to appear only briefly, perhaps to deliver a comforting message? The message could be just a comforting measure or a pressing message.

You may be awakened by a voice. You may feel a
bump in the night near your bed. With this, you may
have a messenger ghost. This kind of ghost is said to
be attached to a person. Perhaps your ghost is there
to deliver a message? The message could be just a
comforting measure or a pressing message. This
ghost may wish to inform you of how it died; it
could have been a victim of a crime. Maybe its body
is missing and you may wish to seek a psychic to
communicate with the spirit.

Many spirit left their bodies sudden and have
unfinished business. They sometimes seek out the
living to make contact. Not all haunting of people
are bad and when confronted with such haunting;
sometimes just listening will end the haunting.

Cryptozoology

Cryptozoology is the study of living animals supposedly found outside of their normal range largely due to the lacking of any evidence of their existence, such as Mothman, Bigfoot and the Loch Ness Monster.

The **Loch Ness Monster** is a cryptid that is reputed to inhabit Loch Ness in the Scottish Highlands. It is similar to other lake monsters in Scotland and elsewhere, though its description varies from one account to the next. Popular interest and belief in the animal has varied since it was brought to the world's attention in 1933. Evidence of its existence is minimal and much-disputed photographs.

The Loch Ness Monster

Bigfoot, also known as Sasquatch, is an ape-like cryptid that is said to inhabit forests in parts of North America. A few scientists have expressed interest and belief in the creature by expressing the opinion that evidence collected of alleged Bigfoot encounters warrants further evaluation and testing. Bigfoot is described in reports as a large hairy ape-like creature, in a range of 6–10 feet tall, weighing in excess of 500 pounds, and covered in dark brown or dark reddish hair. Alleged witnesses have described large eyes, a pronounced brow ridge, and a large, low-set forehead; the top of the head has been described as rounded and crested. Bigfoot is commonly reported to have a strong, unpleasant smell by those who claim to have encountered it. The enormous footprints for which it is named have been as large as 24 inches long and 8 inches wide Bigfoot remains one of the more famous examples of a cryptid within cryptozoology.

Skeptics of Cryptozoology

Skeptics of cryptozoology should note, native pygmies of the Congo reported for years that a creature called the Okapi that looked like a cross between a zebra and a giraffe was in the Congo region. It wasn't until 1900 that the explorer, Sir Harry Johnson, was able to send to Europe a complete skin and two skulls of the creature proving its existence.

The **Okapi** was first sent to zoos in 1920. They can now be found in captivity around the world, though they are rare in the wild. Scientists believe they are the closest living relative of the giraffe. If nothing else the lesson from the Okapi is it is important for naturalists to take serious the stories of native peoples and reports from cryptozoologists.

Okapi

Crop Circles

Crop circles are often large patterns created by the flattening of wheat, corn, etc. Crop circles are also referred to as crop formations, because they are not always circular in shape. The documented cases of crop circles have substantially increased from the 1970s to the present. Ninety percent of crop circles are located in southern England. Many of the formations appearing in that area are positioned near ancient monuments, such as Stonehenge.

There are many reports of UFO sightings and circular formations in swamp reeds and sugar cane fields. The most famous case is the 1966 Tully "saucer nest," when a farmer said he witnessed a saucer-shaped craft rise 30 or 40 feet up from a swamp and then fly away.

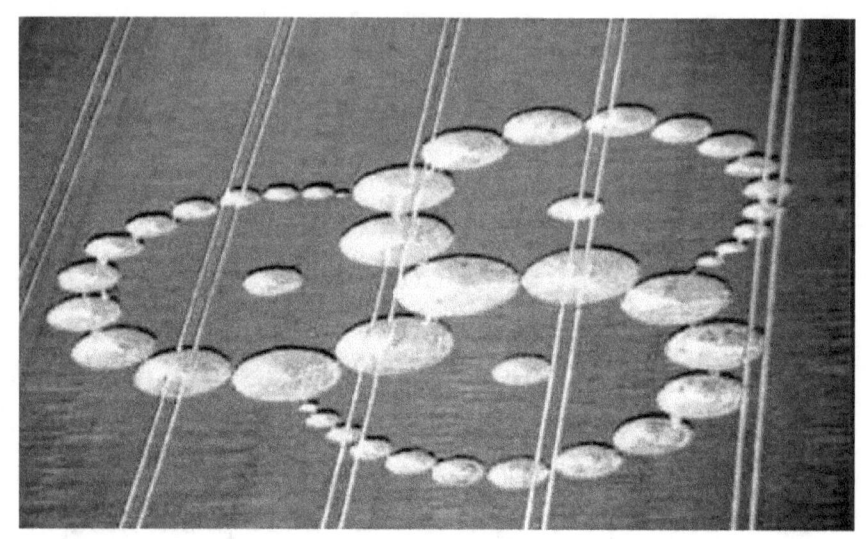

Crop Circle

When he went to investigate the location where he thought the saucer had landed, he found a nearly circular area 32 feet long by 25 feet wide. With widespread media attention in the 1980s, crop circles became the subject of speculation by various paranormal, ufological investigators ranging from proposals that they were created by bizarre meteorological phenomena to messages from extraterrestrial.

Crop Circle Researchers have investigated strange electromagnetic fields associated with the crop circle phenomenon. With these reports, UFO sightings and the meteorological phenomena being associated with crop circles, it doesn't appear that answers to what truly causes crop circle are soon to be know.

UFO's/USO's

An unidentified flying object, often abbreviated UFO, is an unusual apparent observance of an object in the sky that is not readily identifiable to the observer as any known object. The claim of an observance of a UFO is not the admission of a belief in Extraterrestrial life.

While technically a UFO refers to any unidentified flying object, in modern popular culture the term UFO has generally become synonymous with alien spacecraft; however, the term Extraterrestrial Vehicle is sometimes used to separate this explanation of UFOs from totally earthbound explanations.

Some observers argue that because these objects appear to be technological and not natural phenomena and are alleged to display flight characteristics or have shapes seemingly unknown to conventional technology; the conclusion is that they must not be from Earth.

Not to dismiss UFO's but less is known about USO's or unidentified submerged object. A USO is defined as any object or mechanical detection phenomenon of unknown origin observed under water that remains unidentified even after thorough investigation. The maritime analog to UFOs, or unidentified flying objects, unidentified submerged objects is often seen by those who study unidentified flying objects as a related phenomenon. Many sightings of USOs constituting "typical" UFOs emerging from the water are seen as a subset of the latter rather than of the former. In this chapter, I will focus on USO's as to give you a better understanding of this less known subject.

I've always been fascinated by the sea, my parents moved to Myrtle Beach when I was in High School and I spent my teen years walking the beach and looking out over the horizon into the mystery of the sea. I've often wondered what was out there or under there.

Throughout UFO/USO History, there have been reports of undersea alien bases of operation. Do I think such a thing exists or has ever existed?

I do not know but in my own observations from living near the sea for years and hearing reports, I do think there is something to the sightings. I think personally, there could be some truth to the concept of "undersea/underwater" bases of operation. However, whether or not one can determine the origins of these so-called bases is another matter.

My personal observations have only left me with more questions then answers and I see no evidence to suggest a concluded investigation in the near future. I do think whether it be UFO's or USO's that there will be concrete evidence in the future of Alien contact and I also feel it will not be left in mystery but left in the center of some major city of the world.

Skeptics find the need to ridicule and search for factual information in most of the skeptical commentary on UFO's and cryptozoology and other areas of the paranormal and one is usually left with nothing. This is not surprising. After all, how can one rationally object to a call for scientific examination of evidence?" Skeptics, who flatly deny the existence of any unexplained phenomenon in the name of 'rationalism,' are among the primary contributors to the rejection of science by the public. People are not stupid and they know very well when they have seen something out of the ordinary. When a so-called expert tells them the object must have been the moon, a mirage or their imagination, he is really teaching the public that science is impotent or unwilling to pursue the study of the unknown. One must remember that it was a scientist who first concluded the world was flat, built chemical weapons, the atomic bomb, so how can you look for a rational discussion in such a community?

I don't mean that as an insult to the scientific community but I do feel that some in the scientific community have the idea that only those things that we can see, feel and touch exist and nothing else. This to me is irrational in many ways. UFO's are today in the same category as was the Okapi in the 1800's. Will we someday find another Sir Harry Johnson of the ufology community who delivers to a skeptical scientific community, a being from another world? When this day comes, will our so call scientific experts deny the beings very existence to its face or will they simply declare war?

Police and Paranormal

I'm often ask about my prior career in law enforcement but I can't comment on cases I was involved with. I can say that I worked on numerous cases involving devil worship, cattle mutilations and yes, UFO reports. I was even called to investigate a burglary in a haunted house.

I never really saw myself as leaving law enforcement. I also never really saw myself as staying in the career. There are pluses and minuses in all careers but for me, law enforcement was just a job and not something I had a desire to follow to retirement. I worked for several agencies, both private and government. The state was the final agency I worked for, but in total, my investigation experience was about 20 or so years before taking an early retirement and writing my first book.

I think everyone should be "skeptical" of reports of paranormal activity but not necessarily a skeptic as the label is commonly used. I also think we should be open to believing, but not believe all that we hear until we've done our own homework. I did have "paranormal" type experiences in law enforcement that I could not explain, but I've had many more since then as a paranormal investigator.

My main interest has always been UFOs. Ghost hunting stimulates my psychic ability and along the way I've learn more about my own abilities in dealing with spirits. But I find the thoughts of aliens' visiting very interesting. I guess because I've always seen spirits and every other type of phenomenon but UFO's are less common.

I think most police officers believe in the paranormal. I know my father who was a thirty year veteran police officer was very open to the paranormal world. I've worked with numerous officers who had a deep desire to see a UFO during night patrol.

Most officers are skeptics of reports of paranormal activity but being a skeptic is part of the job, it doesn't mean you would not be believed if you reported the activity. As with police officers, most paranormal investigators are skeptic until an investigation is complete.

Anyone performing any kind of an investigation should be open to the possibility that there could be another explanation. This is simply good investigating practice.

Police chase UFO over Cardiff, UK

Cattle mutilation

Cattle mutilation was one of my first cases as a law enforcement investigator. The case involved a Black Angus cow that was mutilated in the middle of an intersection on a mountain road in Western North Carolina. The cow was in the center of a pentagram drawn with red spray pant. Obviously this was satanic worship but not all cases of cattle mutilation are so obvious. Cattle mutilation is the apparent killing and mutilation of cattle under unusual circumstances.

Sheep and horses have allegedly been mutilated under similar circumstances. A hallmark of these incidents is the surgical nature of the mutilation, and unexplained phenomena such as the complete draining of the animal's blood, loss of internal organs with no obvious point of entry, and surgically precise removal of the reproductive organs and anal coring.

Another reported event is that the animal is found dumped in an area where there are no marks or tracks leading to or from the carcass, even when it is found in soft ground or mud. The surgical-type wounds tend to be cauterized by an intense heat and made by very sharp/precise instruments, with no bleeding evident. Often flesh will be removed to the bone in an exact manner, consistent across cases, such as removal of flesh from around the jaw (Image 1) exposing the mandible.

Another odd aspect of the cattle mutilation phenomenon is veterinarians and cattle mutilation investigators claims that mutilated cattle are avoided by large scavengers "such as coyotes, wolves, dogs and bobcats" for several days after its death. Similarly, domestic animals are also reported to be "visibly agitated" and "fearful" of the carcass.

When know human explanation can be found such as in the case in Western North Carolina, how do we explain cattle mutilation? Obviously it becomes paranormal.

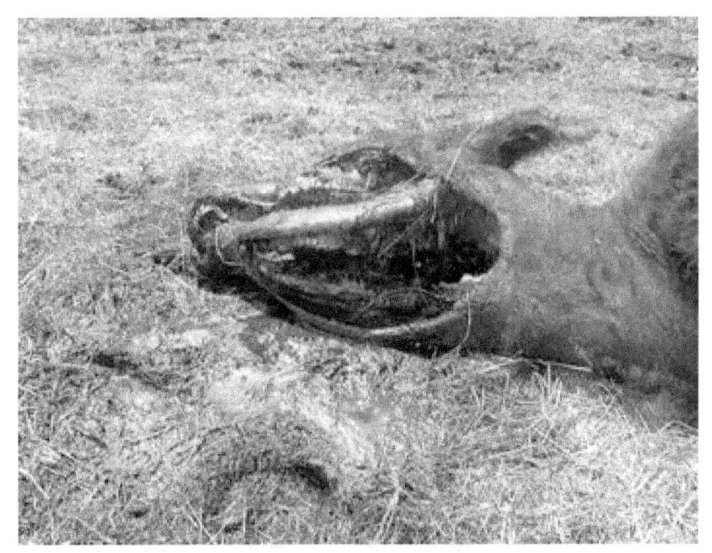

Cattle Mutilation

Most paranormal investigators I've spoken with have suggested that cattle mutilations have been committed by aliens gathering genetic material for unknown purposes. Most of these conclusions are based on the premise that earthly entities could not perform such clean dissections in such a short space of time without being seen or leaving evidence behind at the mutilation site, suggesting the use of unconventional cutting tools in relation to the site.

Some suggest that as cows make up a significant part of the global human diet a study is being carried out on this element of the human food chain. Numerous speculative theories abound, but others center on possible specific nutrient requisites, hormone procurement, species propagation (reproduction), and rote experimentation on mammalian populations.

Conclusion

As a former law enforcement officer, I see the paranormal investigative field as taking over where conventional methods stop. I've had people contact me about missing person cases because the local authorities are left without clues to a case. This is when I used clairvoyance to attempt to secure information.

In UFO cases, paranormal investigators take over when the military and local authorities all seem to have no answers or give unreasonable answers such as swamp gas, stars or simply refuse to comment.

Most paranormal investigators or teams that I work with are trained in investigation techniques either by prior military service or law enforcement. Yes, there are unqualified investigators and those seeking help should as about an investigators experience and qualifications.

Most investigate all kinds of unexplained paranormal phenomena and are devoted to their clients needs. They achieve successful results for their clients through a combined use of research, scientific equipment and personal sensitivity to the surroundings. They do not use Ouija boards, conduct séances, cast spells, or use other "mystical" means to conjure, summon or manifest spirits. Their approach to every investigation is done in accordance with scientific principles, and in a respectful and professional manner. All information collected is analyzed carefully and kept private. I feel that paranormal investigations do not conflict with religious beliefs because the investigator is just seeking answers to what human kind does not understand. Paranormal means out of the normal (unexplained) and the first word in UFO is unidentified or (unexplained). As human beings, we seek answers, it how we learn. To not seek answers and have no desire for answers is truly unnatural and unexplainable.

Paranormal Photo Gallery

UFO over New Zealand

UFO over Scotland

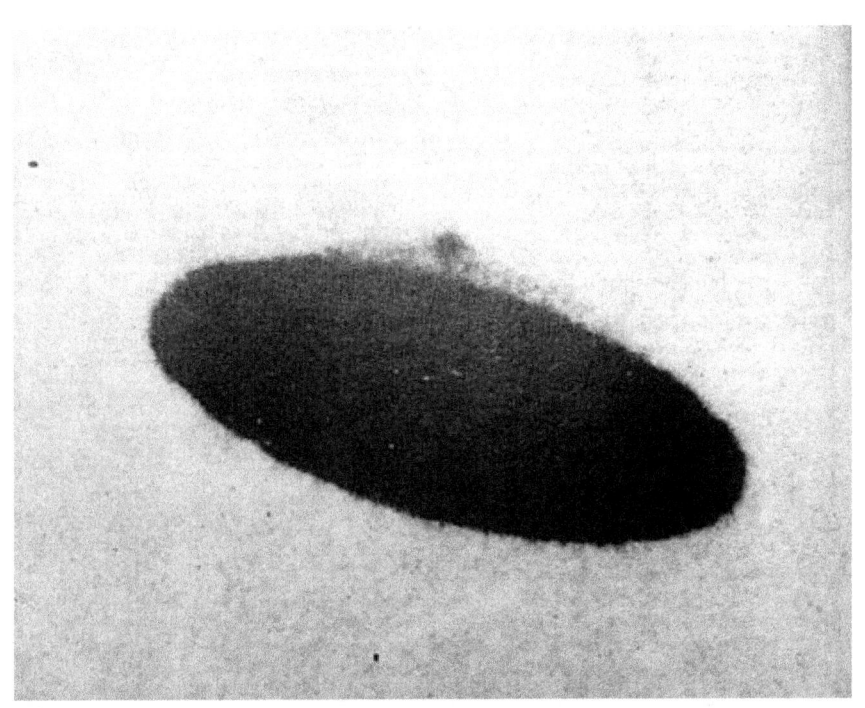

UFO over Arizona

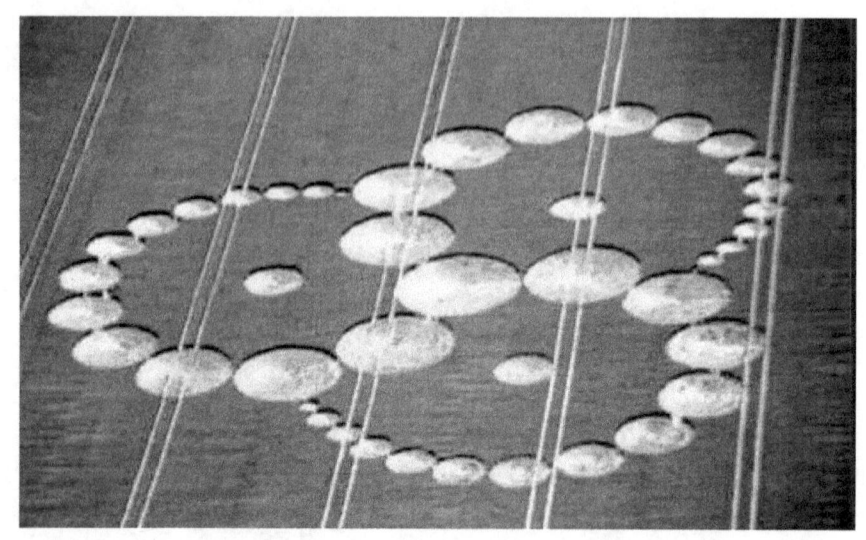

Crop Circle in England

The Brown Lady

Lock Ness Monster

Police case UFO

School house ghost janitor

To all those who have a desire to believe, Thank you!

To my wife, sons, sisters and brothers, I love you!

To those who show me respect, I respect you!

- Rich Arrington

www.ingramcontent.com/pod-product-compliance
Lightning Source LLC
Chambersburg PA
CBHW071631170526
45166CB00003B/1283